LIVING TOGETHER IN NATURE

Jane E. Hartman

LIVING TOGETHER IN NATURE

How Symbiosis Works

drawings by Lorence F. Bjorklund

Holiday House · New York

Text copyright © 1977 by Jane E. Hartman
Illustrations copyright © 1977 by Lorence F. Bjorklund
All rights reserved
Printed in the United States of America

Library of Congress Cataloging in Publication Data

Hartman, Jane E
  Living together in nature.

  SUMMARY: Discusses mutually beneficial relationships
between certain pairs of animals and plants.
  1. Symbiosis—Juvenile literature. [1. Symbiosis]
I. Bjorklund, Lorence F.   II. Title.
QH548.H28   574.5′24   77-3241
ISBN 0-8234-0303-3

FOR JEAN

Certain animals live with each other, or with plants. Some actually need each other to stay alive. Alone, one might not be able to find food, or shelter, or even protect itself from attack. Together with its companion, however, it gets these things it needs.

In most cases, both companions get what they need without harming each other. Sometimes one animal gains more than the other. Some animals and plants, called parasites, simply injure the other in getting what they need.

We call all these companionships in nature *symbiosis*. This means "living together."

If you look in a garden in spring and summer, you may find ant herdsmen tending their "ant cows," or aphids, as carefully as a farmer takes care of his dairy cattle. Aphids are very small insects that feed on plants such as roses. When the ants stroke them with their feelers, the aphids give off a sweet liquid called honeydew that the ants like very much.

In turn, the ants protect the aphids from their enemies, and move them to choice feeding areas. They even take the aphids and their eggs to underground shelters for the winter.

In Mexico there are ants that herd caterpillars. All day the ants keep the caterpillars in covered burrows. At night they lead them out to feed on a flowering bush, and guard them while they eat. Scientists have found that if the caterpillars were allowed to feed during the day, they would all be eaten by their enemies.

The caterpillars, like the aphids, provide sweet honey-dew which the ants drink. After eighty-three days of care by the ants, the caterpillars turn into rare butterflies and fly away.

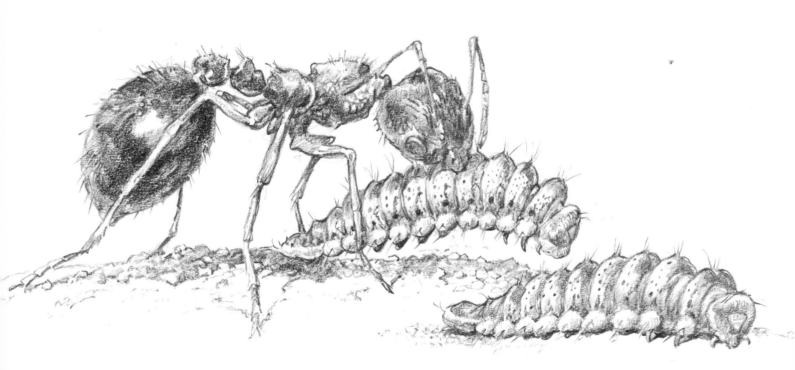

The tickbird is a small African songbird that clings to the tough hide of a rhinoceros. It helps to get rid of ticks and other annoying pests on the rhino's skin by picking them off and eating them. The sharp-eyed bird also warns its large companion of attackers approaching. In turn the tickbird gets food, and also protection. What animal would dare attack a bird on the back of a rhino?

A bird with unusual tastes is called the honey guide. It loves beeswax but can't open beehives to get at it. When the bird finds a hive, it cries excitedly and searches about for a partner to help. The bird's favorite companion is the ratel, whose long claws and thick protective coat make it an ideal beehive robber. The ratel follows the honey guide until it sees the beehive. Then it tears the hive open with its claws and eats the honey. When the ratel is through, the honey guide eats all the beeswax it wants.

Although the honey guide loves to eat beeswax, it cannot digest it. But there are very small organisms called bacteria in its intestines, and these bacteria break down the wax so the bird's body can use it. Useful organisms like these bacteria have a protected home in the digestive tracts of many animals, including humans.

On the plains of Africa, the zebra and the long-necked ostrich
need each other's help to guard against their enemy, the lion.
The ostrich's sharp sight, combined with the zebra's keen
senses of smell and hearing, can discover a lion soon enough
to allow both creatures to escape without harm. Alone, neither
would be able to locate the danger as quickly. So zebras and
ostriches team up for their mutual protection.

The crocodile is an animal that likes to sun itself on the banks of streams. Its partner, the Egyptian plover, is a small shore bird. As the crocodile lies sunbathing with its mouth open, the plover eats blood-sucking pests, called leeches, off its body and out of its mouth. Fearlessly the bird hops about picking up leeches even between the sharp teeth of the crocodile. If danger threatens, the bird warns the crocodile in time for it to slide into the water.

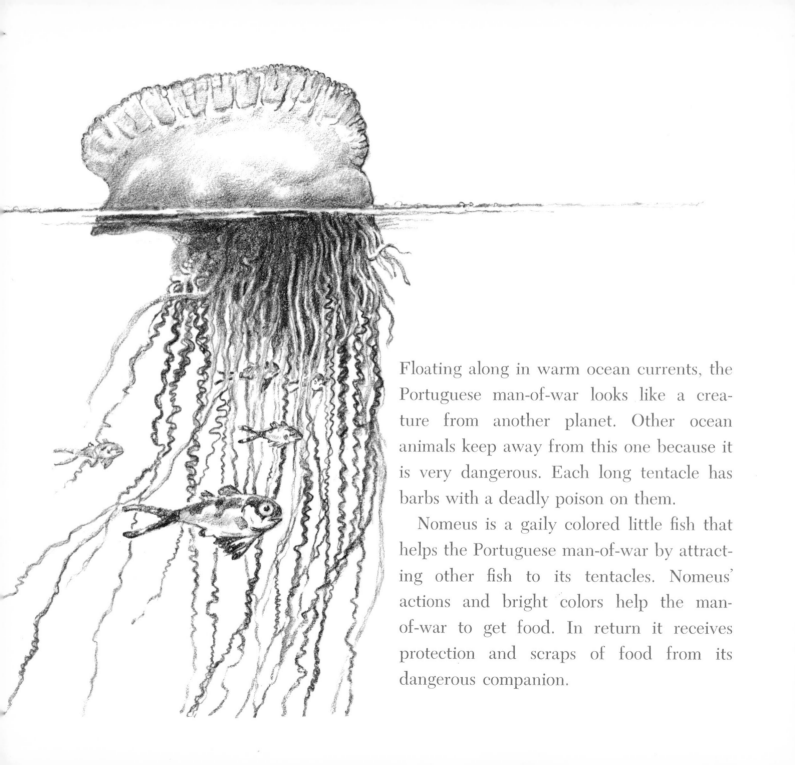

Floating along in warm ocean currents, the Portuguese man-of-war looks like a creature from another planet. Other ocean animals keep away from this one because it is very dangerous. Each long tentacle has barbs with a deadly poison on them.

Nomeus is a gaily colored little fish that helps the Portuguese man-of-war by attracting other fish to its tentacles. Nomeus' actions and bright colors help the man-of-war to get food. In return it receives protection and scraps of food from its dangerous companion.

The moray eel is a ferocious creature that eats shrimp as part of its diet. But there is one type of shrimp the eel will not hurt. This small red and white cleaner shrimp sits on a coral and waves its claws and antennae. This attracts the attention of the eel, which swims in where the shrimp can take hold of its body. Then the shrimp eats parasites off the eel's body, gills, and mouth. When the job of cleaning the eel is finished, the shrimp lets go and the eel swims away.

SHIELD LICHEN
ON A TREE

MUSHROOMS,
A KIND OF FUNGUS

Fungi are plants, such as molds and mushrooms. Algae are green plants that need light to make food. When both of these plants get together they form a plant called a lichen. These come in many forms. They are very tough and can live almost anywhere.

Although both algae and fungi can live by themselves, they help each other when they join forces. The fungi may protect the algae. And they absorb and store water and other materials the algae need. In turn their partners supply them with food.

Mistletoe is a parasite plant. It sends small rootlike threads into its host plant, usually a tree, and sucks out water and minerals. Under some conditions, if the mistletoe uses too much water, the tree may die. When this happens, the mistletoe dies too.

The yucca moth depends on the sweet white blooms of the giant yucca plant. The female moth chooses a fresh flower and gathers pollen. Then she flies to another flower where she deposits her eggs deep within the bloom. At the same time she spreads the pollen she is carrying.

When the moth eggs hatch, the infant moths, or larvae, are inside a yucca fruit, which is full of seeds. Here they feed on the seeds while they grow.

In the Sonoran desert of southern Arizona, the Gila woodpecker drills a hole in a giant saguaro cactus. The saguaro seals off the injury with a substance something like cork. This makes a fine weatherproof nest hole for the woodpecker. When it moves in, the bird eats moths that harm or destroy the cactus. Both the plant and the bird gain from this companionship.

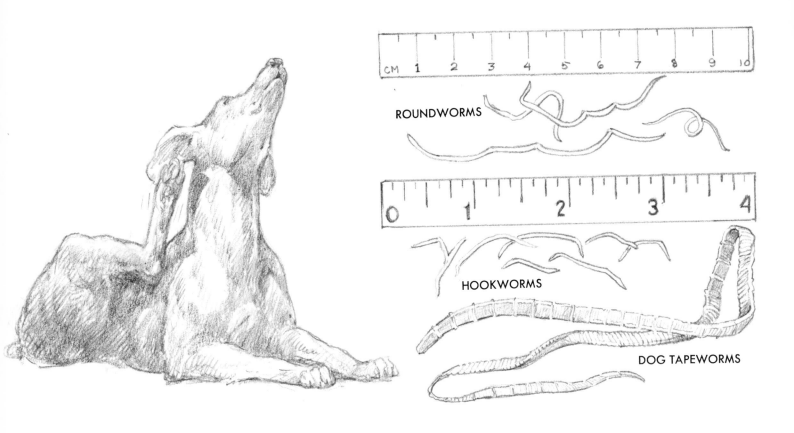

ROUNDWORMS

HOOKWORMS

DOG TAPEWORMS

Some companionships benefit only one animal or plant, while causing harm to the other. The one that gains is called a parasite. The other is the host.

Fleas that may live on cats and dogs are parasites. They suck blood from them, causing their skin to itch. Many fleas

22

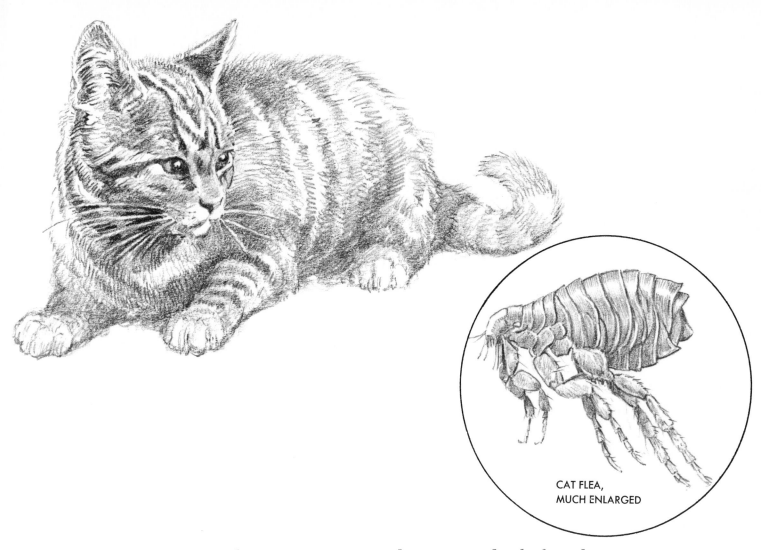

CAT FLEA,
MUCH ENLARGED

on an animal can cause great weakness, even death, from loss of blood. Fleas can also carry serious diseases.

Worms too can cause serious injury to a host. Hookworms, roundworms, and tapeworms are found in humans and other animals.

Some ants grow a special kind of fungus in their underground homes. They gather leaves and chew them into pulp. This is used as soil for the fungus to grow on. The fungus could not live without the ants' care. The ants weed and fertilize their fungus gardens, so the crop does well and supplies the ant colony with food.

In many ant colonies there are "guests." The beetle called Lomechusa is one. The ants take the beetle's honeydew, and they feed, groom, and protect the insect even though it may occasionally eat a young ant.

It is fortunate for the ants that only a few beetle young survive the frequent turning and handling they must endure from their ant nurses. If they lived, the beetles would soon overrun and destroy the ant colony.

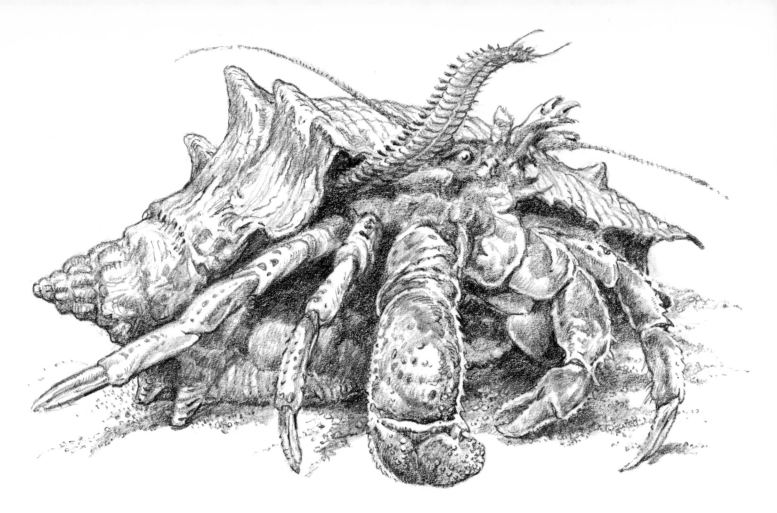

If you have spent time at the seashore, you may have seen a hermit crab. This crab backs itself into an empty shell and carries it on its body as its home. The crab often has a bristleworm as a "boarder" in the shell. The bristleworm shares food and lodging with the crab. In turn, the worm keeps their house clean by eating any food scraps that might remain in the shell and decay.

Melia is a crab that carries a small sea animal, called an anemone, in each front claw. As a very young crab, Melia grasps the stinging anemones and never lets go of them as long as it lives.

The many-tentacled anemones help protect the crab from attackers. They also help catch food, by stinging prey and making it helpless. In turn, the crab carries the anemones to places they could never go, where there may be more food. Anemones are stationary creatures that must attach themselves to other objects, or to the ocean floor.

The sea anemone has other strange companions. The pretty damselfish lures other fish into the stinging tentacles of the giant anemone. In turn, it receives protection, and is never stung by the anemone. In fact, the damselfish can even find shelter in the stomach of the anemone without fear of being poisoned. Besides sharing food, the damselfish, using its fins and tail, fans water around the anemone, which helps it to breathe better.

In the warm waters of the world's seas, a fish called a remora, or "shark sucker," sticks to larger fish like sharks, marlin, or swordfish. The remora has a sucking disc on the top of its head. The fish attaches this to the large fish's body, without harming it at all. In this way these hitchhikers of the sea get free transportation. They also receive scraps of food and protection from the large fish.

Fish of all kinds crowd the cleaning stations at a coral reef in tropical seas. They are waiting their turns to be cleaned by the wrasses, or doctor fish. The brightly colored doctor fish groom deadly moray eels and scorpion fish without fear of attack. They eat off parasites. They clean sores and swellings. The doctor fish work inside the mouths and gills of the fish as well as on their scales. The doctor fish get food and freedom from attack as they keep other fish healthy.

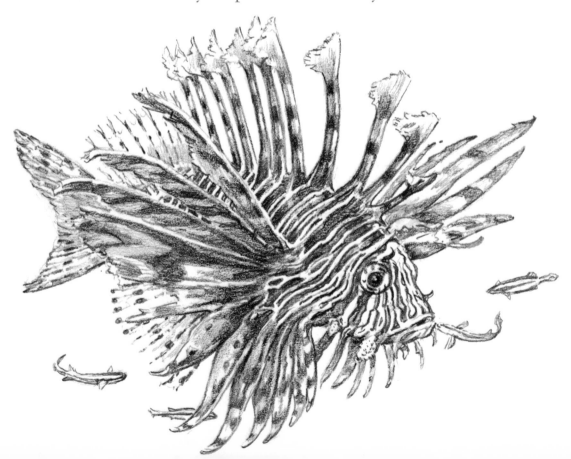

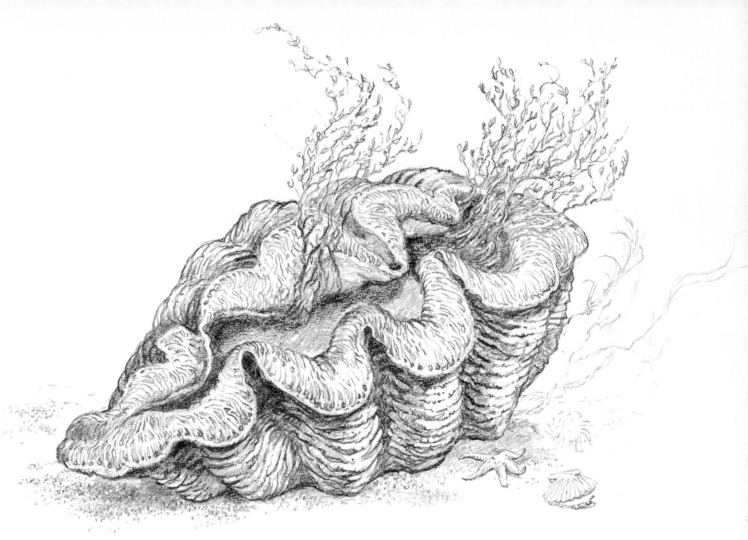

In the shallow waters of a South Pacific coral reef, the giant clam lies with its two shells open. In this way the algae growing inside its shell get light. These simple plants get water, carbon dioxide, and nitrogen from the clam. They use these materials to manufacture food. The clam actually has a self-contained farm. When it gets hungry it eats some of the algae.

The sooty shearwater is a graceful gray sea bird that lives in burrows on islands off the coast of New Zealand. During the day the bird spends its time fishing, leaving its burrow empty except for its young.

Another resident of these islands, the tuatara lizard, likes to sleep in a burrow, too—the sooty shearwater's home. The tuatara sleeps all day and hunts for insects at night while the sooty shearwater sleeps. The arrangement keeps their home free of insects. Neither animal bothers the young of the other. The tuatara benefits more from this partnership—it doesn't have to dig a burrow.

Some African birds take advantage of the burrow dug by the African aardvark. They fly into the aardvark's domed entrance and dig a pocket-like room in the roof of the tunnel. In this sheltered spot they nest and raise their families. It makes no difference to the birds whether the aardvark is living there or not.

The cattle egret is often seen in fields following grazing animals and looking for insects stirred up by their feet. It also rides on their backs, eating pest insects from the animals' skins.

The fish hawk, or osprey, builds a huge platform nest of sticks high in a tree. Each year she adds to her nest, making it even larger. Other birds, such as the black-crowned night heron, find this nest attractive because it is safe. Occasionally they even find food scraps left by the bigger bird.

The osprey's home is like an apartment house. The osprey lives on the top while the rest of her tenants live among the sticks below.

In the lush green forests of South America, the three-toed
sloth is partly protected from its enemies by green algae
growing along its stiff hairs. These algae are very small plants
that need light to make food for themselves.

The slow-moving sloth lives its upside-down life high in
the trees where there is bright light for the algae on its fur. In
turn, the algae give the sloth a somewhat greenish color, and
so it blends better with the leaves in the trees where it lives.

The animals and plants that come together in symbiosis find
a better way of life. Once a useful companionship has formed,
their descendants keep on doing the same thing. Then we can
see that symbiosis has appeared once more as a way of living.

INDEX